GRAPHING PRIMER

Laura Duncan Choate
JoAnn King Okey

DALE SEYMOUR PUBLICATIONS

Cover designer: Rachel Gage
Editor: Frances Christie

Order number DS01904
ISBN 0-86651-486-4

DALE
SEYMOUR
PUBLICATIONS
P.O. BOX 10888
PALO ALTO, CA 94303

abcdefghi-MA-895432109

Contents

Introduction

Graphing Primer is part of a series of materials from Dale Seymour Publications that is designed to help you bring graphing into your primary classroom. This volume provides an extensive, practical introduction to graphing techniques; it is full of ideas on how to create a wide variety of enjoyable group and individual graphing experiences. Included are reproducible blackline masters to help students collect and record data and learn to read, analyze, and interpret graphs.

You will also find suggestions on how to make the best use of two other items in this primary graphing series: *Graphing Grids* and *Pictograms: Graphing Pictures for a Reusable Classroom Grid*. These items, available separately, are designed especially to complement *Graphing Primer*. *Graphing Grids* is a set of four reusable 34-by-22-inch blank grids for making both floor graphs and wall graphs. The *Pictograms* book contains nearly 400 pictures to duplicate and use with the grids, along with ideas for topics that you can graph using the pictures.

WHY GRAPH?

Graphing is an ideal math activity at the primary level. It allows young students to relate to numbers both tactilely and visually. The reading of words is reduced to a minimum. Graphs can show at a glance what could take paragraphs to describe. As a way of gathering and looking at real-world data, graphing helps students relate math to their personal experiences.

Graphs provide a structure for organizing and simplifying collections of figures. This is likely your students' first experience with statistics. As they gather and graph information in the classroom and make comparisons of the data, students discover that statistics are an integral part of daily life, and that such statistics can help them draw mathematical conclusions about the world around them.

WHAT SKILLS DOES GRAPHING INVOLVE?

Graphing everyday data develops a number of different skills. At the most elementary level, students learn to identify similarities and differences. Gradually they become more sophisticated in defining the attributes of different objects. Graphing develops the ability to sort, to classify, and to place objects in one-to-one correspondence on the graph.

Making the graph itself is only the first step. Students also develop skills in interpreting graphs. They must be able to count the amounts in each category and do simple visual comparisons. Addition and subtraction skills are used as students begin to do exact comparisons. Problem-solving skills are called into play as students try to determine why certain results were obtained, whether or not those results could be generalized to a broader situation, and if the results might be used to predict future situations.

ALWAYS BE READY TO GRAPH

Opportunities for graphing occur daily in the classroom, yet we often let them slip by, thinking that we simply don't have time to prepare everything needed to make a graph. That's where this series of materials can help out. With a set of reusable graphing grids, a collection of pictograms, and the ideas in this *Graphing Primer,* you can be prepared to take advantage of any graphing opportunity that comes along. The more frequently you do graphing in the classroom, the more adept your students will become at interpreting graphs and developing their critical thinking skills.

WHO CAN BENEFIT FROM *GRAPHING PRIMER?*

Primary classes. When working with the whole class or smaller groups, you can copy graphs and blackline masters for use on an overhead projector.

Individual students. Make copies of blackline masters for individual student use to provide reinforcement and practice of graphing skills.

Beginning readers. The graphs in *Graphing Primer*, developed with pictures and symbols, require little or no reading skill.

Fast workers. Students who finish other work quickly can use many of these activities and blackline masters at classroom centers.

Gifted students. Many of these activities provide application of higher-level thinking skills and can be extensions of simpler investigations.

Other ages. Although we suggest using this series with kindergarten through second grade students, pre-schoolers and older students alike can also benefit from the graphing experiences detailed in these pages.

Substitute teachers. If substituting in a classroom, you will find these graphing activities easy to apply to any teaching situation. Graphing experiences are a good way to become acquainted with students and to actively involve all students in a group lesson.

TEACHING THE THREE STAGES OF GRAPHING

Students are more likely to understand graphing if we present it in three developmental stages, progressing from the concrete to the abstract.

Real graphs. At the first stage, students make *real* or *object* graphs, comparing actual objects that they place in the squares of a blank grid laid flat on a table or floor. They start with a real situation, such as "The Types of Shoes We Wear," sort real objects (the shoes) by attribute (buckles, laces, Velcro, slip-ons, boots, and so forth), and place the shoes in categories on a large blank grid to form a real graph. This concrete stage lays the foundation for all other graphing activities.

NOTE: The blank *Graphing Grids* are suitable for graphing real objects that will fit in 3-inch squares. To graph such objects as shoes, rule a larger grid on posterboard.

A REAL GRAPH USING REAL OBJECTS FOR STUDENT RESPONSES

Picture graphs. At this intermediate stage, students graph with pictures—graphic symbols of something real—rather than with real objects. This stage forms an important bridge between real and abstract graphs.

A PICTURE GRAPH
USING PICTURES FOR
STUDENT RESPONSES

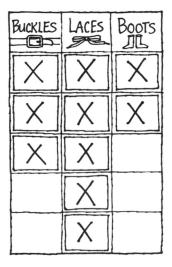

AN ABSTRACT GRAPH
USING X'S FOR
STUDENT RESPONSES

Abstract graphs. In this third stage, students use symbols—such as X's, circles, or squares—to represent real things. This is the most abstract level since, to find meaning, students must relate each symbol back to the real object it stands for.

Beginning graphers should have the chance to graph the same set of data sequentially through all three stages of graphing. Exposing students to the same data in different forms will enhance their understanding of the relationship between real (concrete) and abstract (symbolic) graphs. Repeated experiences going from concrete to abstract help students develop the ability to generalize and apply the problem-solving skills used in interpreting graphs to other situations. Students should have many experiences with group graphs before creating individual graphs. The Group Graphing Experiences section of this book gives more information on group activities for all three stages of graphing.

TALKING ABOUT YOUR GRAPHS

Encourage students to discuss and interpret every graph they make. Before making a graph, discuss with the students what questions they want the graph to answer. Ask students to predict what the graph might show, for instance: "Which flavor ice cream do you think will be most people's favorite?"

Once the students complete the graph, ask specific questions that lead them to react to the graphed data. Possible questions to ask about graphs are:

What does this graph tell us?
Are any columns (or rows) the same?
Which column (row) has more?
Which column (row) has fewer?
Which column (row) has the fewest?
Which column (row) has the most?
How many ——— are there?
Are there more ——— or more ———?
How many more ——— than ——— are there?
Are there fewer ——— or fewer ———?
How many fewer ——— than ——— are there?
How many ——— are there altogether?
Tell us a number sentence about the graph.
What else can we say about this graph?
What other way could we have organized this graph?
What other information might we want to know?
What questions did we ask when we started this graph?
If we gave another class the same graph to complete, would the results be the same?

Students can later translate the simple statements that they made about a graph into symbolic math language. That is, "Four people chose vanilla ice cream and three people chose chocolate ice cream" becomes 4 + 3 = 7.

After discussing a graph, you might leave it on display for several days, allowing students to make additional discoveries on their own and perceive relationships that they may not have noticed previously. As students become more comfortable with graphing, you will find them asking questions and discussing the graphs together during free time, even without your direction. Students might discover that a graph does not always provide the answer to every question they ask. Group discussion of classroom graphs prepares students to answer questions about the individual graphs they will be making as they become more proficient.

ABOUT THIS BOOK

This book is organized into two broad categories: group graphing experiences and individual graphing experiences. Group graphing experiences include creating classroom and school-wide graphs, using finger puppets to graph student opinions, and conducting group surveys to collect data for graphing. Individual graphing experiences include sorting materials in graph-it bags, counting and graphing items in pictures in graphing folders, taking surveys for individual polling graphs, and using different ways to gather data from home. The individual experiences are more sophisticated and require more independent application; students should try them only after many group graphing experiences.

The description of each particular graphing technique is followed by blackline masters for using that technique. At the end of the book are two Graphing Expert certificates that you can duplicate and award to students when they have successfully completed a range of the activities in this book.

Group Graphing Experiences

CLASSROOM GRAPHS

Group graphing experiences are an ideal way to become better acquainted with students and to actively involve all students in a group lesson. Once you understand the basic approach, you can pull together a graph at a moment's notice. Later in this section you will see how to set up a permanent graphing display that can be used over and over for classroom graphs.

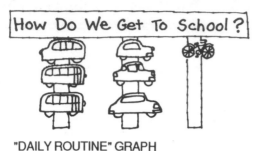

"DAILY ROUTINE" GRAPH

"LONG-TERM" GRAPH

When Can Group Graphing Experiences Take Place?

All day every day. You can integrate graphing experiences throughout the entire school day because it is practical to graph information about all curricular areas (such as science, literature, and social studies). Display a graph only for one day, or leave it up on display for several days or weeks, allowing students time to make further discoveries as they interpret the results.

Over the long term. Some graphing activities are cumulative in nature and can be ongoing, year-long projects. For instance, students collect data over time as they record the number of teeth lost each month, whose birthday is when, and what the weather is like on a daily basis.

In your daily routine. You can effectively use graphing techniques in certain areas of classroom management. This is an excellent chance for children to see that graphs are a practical real-life tool. Here are some elements of the daily class routine that lend themselves to graphing:

- **Attendance.** Students can use names or photographs to sign in on a graph each morning. This is quick and easy in addition to building responsibility and self-reliance.

- **Sharing.** Before class begins each day, students wishing to share information, objects, or personal stories can move their symbols to the positive or negative column on this "Do You Have Something to Share?" graph.

• **Lunch and/or milk.** If you are responsible for gathering information about which students require hot lunch, a daily graph can provide that information quickly. You can also assemble orders for snack or lunch milk from the results of a daily graph.

• **Transportation.** You can display information on how students travel to and from school in graphic form near the exit door. This graph is useful not only to you, but also to any substitute staff.

• **Center choices.** You can use a class graph as a management tool to control the number of students at different activities or centers. List each activity on the graph and highlight the number of available places. Students then use their pictures or names to graph their preferences. When you provide more options than there are students, children can finish one task and then move to another available option.

• **Hall passes.** In some situations students must leave the room while class is in session, perhaps to go to the restroom, to the library, to work with special teachers, or to see the nurse. You can use graphs to track student movement and know where everyone is at a glance. Students needing to leave the classroom would record their movements on the graphs using photographs or nametags. Again you can control the number of available spaces by highlighting.

Ideas for Group Graphs

Comparing self to the entire class. Each year teachers and classes undergo a period of becoming acquainted with one another. This period is also one of learning to recognize and respect similarities and differences within the group. Graphing activities are a perfect way to explore the unique characteristics of each individual and foster self-worth while also building a sense of belonging through shared common experiences. The following topics are ideal for group graphs:

"GETTING ACQUAINTED" GRAPH

birthdays	attendance (boys/girls, present/absent)
ages	type of shoe sole (smooth/bumpy)
hair color	type of shoe worn (boots/sandals/sneakers)
eye color	fingerprint type (whorl/loop/arch)
family pets	first letter of name
shoe size	color of clothing worn (shirt, shoes, socks)
number in family	transportation to school (car/bus/walk/bike)
missing teeth	measurement of hand span
wake-up time	snack brought today
bedtime	glasses/no glasses
weight	type of shoe fastenings (laces/buckles/Velcro)
height	left-handed/right-handed

Choosing favorites. Graphing individual favorites encourages students to express their own ideas as well as to be aware of and respect the ideas of others. Use these graphs also to make group decisions, such as: What game should we play? or Which new song will we learn today? The following topics are ideal for "favorite" graphs:

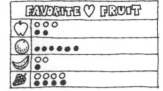

food	color
fruit	number
vegetable	day of the week
pasta	song
cracker	flower
ice cream flavor	friend
snack	toothpaste
lunch	toy
fast food restaurant	ball game
nursery rhyme	sport
fairy tale character	hobby
book character	pet
story	zoo animal
book	dinosaur
television show	school activity
cartoon	classroom job

"FAVORITES" GRAPHS

Making Group Graphs

Pictograms are small pictures that you can use to illustrate the attributes by which you are going to graph. For example, in graphing how we get to school, there could be pictograms of a car, a bike, a bus, and a person walking. Non-readers will find the visual illustrations particularly helpful to define the attributes used in a given graph.

SAMPLE
PICTOGRAMS

This primer includes a number of small pictograms (see pages 13–15). For more examples, refer to *Pictograms: Graphing Pictures for a Reusable Classroom Grid* (from Dale Seymour Publications), which contains nearly 400 pictograms designed for use with the reusable blank *Graphing Grids* (also available from Dale Seymour Publications). Additionally, you and your students can make your own pictograms from photographs, magazine cut-outs, stickers, coloring book cut-outs, or original drawings.

Using pictograms with real graphs. To make a real graph, place reusable grids flat on the floor or table. At this stage of graphing, you might use pictograms to label the categories, mounting each pictogram on a 3-by-5-inch card that is folded in half to stand up like a tent. For example, you could use pictograms of different types of crackers for labels on a real graph of "Favorite Cracker," while students would indicate their responses with real crackers.

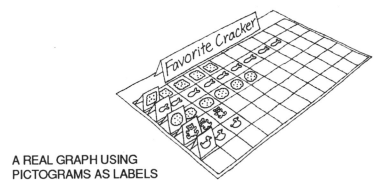

A REAL GRAPH USING
PICTOGRAMS AS LABELS

Using pictograms with picture graphs. You can create a picture graph entirely with pictograms and reusable graphing grids displayed on the wall. For example, consider how you could graph the topic "How We Get To School." You might label the categories with pictograms of a car, a bike, a bus, and a person walking. Students could then mark their responses with identical pictograms that you have duplicated from this book (page 13) or with pictures that the students have drawn themselves.

A PICTURE GRAPH USING IDENTICAL
PICTOGRAMS FOR STUDENT RESPONSES

A PICTURE GRAPH USING PICTOGRAM FACES
FOR STUDENT RESPONSES

Alternatively, they might mark their responses with pictures of themselves. These could be self-portraits drawn on construction paper squares, copies of miniature photographs, or copies of blank-face pictograms that the students have colored to represent themselves. You might collect miniature photographs from cumulative records and duplicate them for this use. Or, photograph three children together and cut the photograph apart to obtain small headshots. In this way, ten exposures would produce portraits for a class of thirty.

Using pictograms with abstract graphs.

For an abstract graph, you would use pictograms only to label the categories, while students record their responses symbolically—maybe with X's written with grease pencil on the laminated grid, or with simple gummed circles, or with students' names written on small squares of paper, such as Post-it notes.

How We Get To School

AN ABSTRACT GRAPH USING
PICTOGRAMS AS LABELS

Using pictograms with tally graphs.

You can also use the pictograms to make a tally graph. Mount them on a large blank sheet of paper for display on a bulletin board or easel, and let students use crayons, markers, or a grease pencil to make a tally mark for each response.

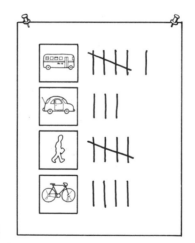

A TALLY GRAPH
USING PICTOGRAMS

When you use pictograms with the blank graphing grids displayed on the wall, attach them in one of the ways shown on page 11 to ensure that you can use your blank grids again and again.

Paper clip mounts. Use a single-sided razor blade or similar sharp instrument to cut along the 1/2-inch slit at the top of each grid square. Slide the tip of a plain paper clip through the slit and push down until the larger side of the clip is behind the grid and the shorter loop is visible from the front.

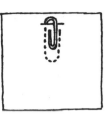

Always keep a paper clip mounted in each square. Then you can easily mount labels and pictograms on the grid by slipping them under the clips.

Paper fastener hooks. Cut the slit at the top of each square as described above. From the back of the grid, slip a 1-inch paper fastener through each slit. Fold the longer section of the fastener flat against the surface of the grid and bend the second section into a hook shape.

To make a graph using this mechanism, simply punch a hole in the top of each label or pictogram and hang it from a paper fastener hook.

Magnet mounts. Tape the grid to a magnetic chalkboard. Use magnets or magnetic tape to mount your labels and pictograms.

Velcro mounts. Secure a small strip of Velcro to the top of each square on the grid and attach a corresponding piece to the back of each label and pictogram.

As labels for categories, pictograms are usually placed in the column of squares farthest to the left or in rows across the bottom or top of the grid. Placing the labels at the right edge (for graphing from right to left) is not recommended at the primary level, since students are still learning the left-to-right reading sequence.

Making a Permanent Graphing Display

Some teachers like to create a permanent display or "graphing center" in their classrooms. This is easily done by mounting a reusable wall grid on a bulletin board under a general title such as "Read Our Graphs" or "We're Speaking Graphically!" While this general title stays in place throughout the year, be sure to leave space on the grid for a specific title, written on a sentence strip, for each new graph you make. Mount the pictogram numerals across the bottom of the grid for making numerical comparisons, and use prepared or student-made pictograms for graphing responses.

When you are finished with the graph, simply remove all the special pictograms and sentence strip title, and your blank grid is ready to use again. Store the pictograms with the sentence-strip title in a labeled envelope or file folder for use at another time.

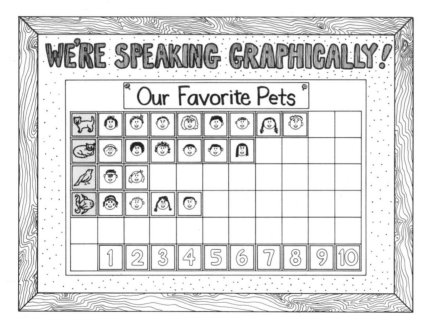

GRAPH DONE AS PART OF A PERMANENT DISPLAY

TRANSPORTATION • FRUIT
How do you come to school?
What is your favorite fruit?

walk	bus	car
bike	grapes	orange
banana	apple	peach

GRAPHING PRIMER **13**

LUNCH • EYES • HAIR
What type of lunch are you having?
What color are your eyes?
What color is your hair?

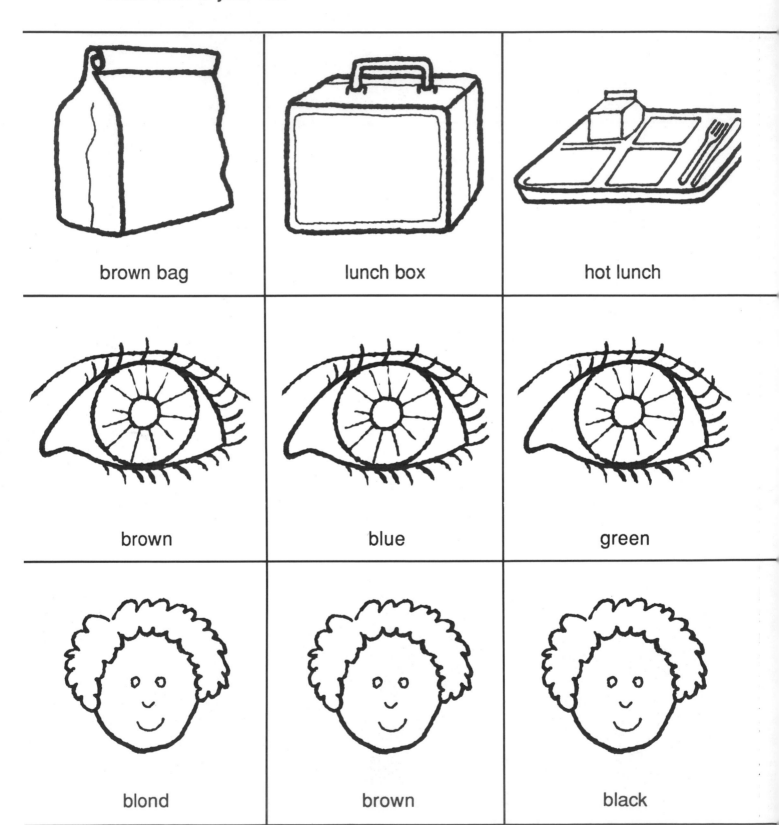

| brown bag | lunch box | hot lunch |

| brown | blue | green |

| blond | brown | black |

FINGERPRINTS • WEATHER
Is your fingerprint a whorl, loop, or arch?
What is today's weather? (Keep track for an entire month.)

whorl	loop	arch
sunny	cloudy	rainy
snowy	foggy	windy

OPINION FINGER PUPPETS

Another type of group graphing experience is the use of finger puppets to graph students' opinions. These puppets are particularly useful for students who might feel uncomfortable sharing their feelings with classmates. The use of puppets sometimes enables students to feel that the puppet is sharing the opinion rather than the student.

Copy and cut out each finger puppet on these pages and tape the tabs together to form a cylinder. When you ask a question, each student holds up the finger puppet that illustrates his or her feeling. The students could then graph the opinions expressed.

Possible questions to use with these finger puppets are:

How do you feel about . . .
    outdoor play?
    taking turns on the equipment?
    jumping rope?
    loud music? quiet music?
    your bedtime?
    eating vegetables?
    taking baths?

    having a babysitter?
    buying lunch at school?
    sharing?
    coming to school?
    riding on the school bus?
    having a loose tooth?
    your best friend?
    wearing the color purple?

Blackline Masters for Opinion Finger Puppets

Cut out figures; tape tabs behind to form a cylinder. Also use flat on a grid as pictograms to graph responses to questions.

happy

sad

angry

no opinion

happy

sad

angry

no opinion

GROUP SURVEYS

Group surveys are another way to approach the group graphing experience. They can provide students with enough data to complete a group graph about their entire class or even the whole school. Use this activity only after students have had many graphing experiences. These graphs, generally done at the picture or abstract level, require more independent application.

Suppose you are graphing a variety of favorites. Duplicate the "My Favorites" survey sheet on page 20 and hand out copies to your students. They write their personal responses in each box on the page, then cut the boxes apart and deposit them appropriately in eight paper bags labeled with the categories. Thus, when the surveys are completed, a class of 30 students would have 30 slips of paper showing their favorite colors inside a paper bag labeled "1. Color." An individual or small group of students would then take the answers to that question, sort the results, and develop a graph.

This section includes blackline masters for two types of survey sheets, "My Favorites" (pages 20–21) and "Getting Acquainted" (pages 22–23). You might display the results of this activity at a parent night. If students have had many graphing experiences prior to this activity, they will form their graphs in creative and individual ways. Students can use these survey sheets for individual graphs, too, and develop their own survey sheets similar to those provided here.

A GROUP SURVEY SHEET AND
RESULTING FAVORITE COLOR GRAPH

SCHOOL-WIDE GRAPHS

Group surveys can lead nicely in to making school-wide graphs. You can extend any of the ideas for classroom graphs to use for this activity. Students could create school-wide graphs on a weekly or monthly basis. A different class might be in charge each time, or a team of students could work with the principal to display the results.

Students from the class in charge should determine what question to ask, hypothesize what the answers will be, and decide how they are going to collect their data. Will they need any special materials? How can they ensure that someone is not polled more than once? Students might collect their data on pages similar to survey sheets, or by keeping tallies as they ask people to answer their question.

Students then organize their results on a graph and display it in a prominent place in the school, such as the library wall, office bulletin board, entryway hall, or lunch room. Having studied the results, students should then be asked to generalize: What conclusions can they draw?

A typical school-wide graph might show in which month most students were born. Another school-wide graph might address a current problem or issue; for example, if the parent organization has $1000 to spend on playground equipment, students could be polled on their choice: swings, basketball court, or geodesic dome climber.

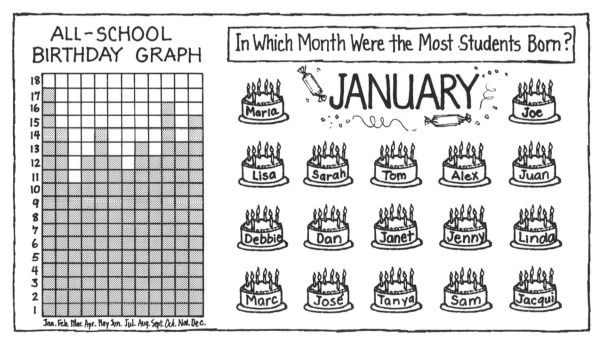

A SCHOOL-WIDE BIRTHDAY GRAPH

MY FAVORITES

1. Color

_____

2. Snack

_____

3. Fruit

_____

4. Holiday

_____

5. Pet

_____

6. Book

_____

7. Number

_____

8. Animal

_____

MY FAVORITES

9. Ice cream

_____

10. Dinner

_____

11. Song

_____

12. School subject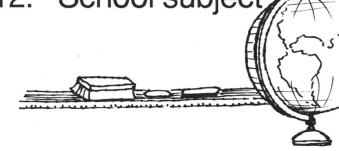

_____

13. TV show

_____

14. Sport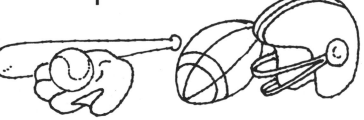

_____

15. Recess activity

_____

16. Toothpaste

_____

GETTING ACQUAINTED

1. Birthday

_____

2. Glasses

YES   NO

3. Eye color

_____

4. Number of pets

_____

5. Hair color

_____

6. Number in family

_____

7. Age

_____

8. First letter of name

_____

GETTING ACQUAINTED

9. Transportation to school

_____

10. Freckles

YES   NO

_____

11. Shoe size

_____

12. Left-handed or right-handed

_____

13. Height

_____

14. Shoe type

_____

15. Weight

_____

16. Number of missing teeth

_____

Individual Graphing Experiences

Individual graphing activities provide experiences at all three levels of graphing: real, picture, and abstract. You can place activities at learning centers to provide individual reinforcement of concepts you have introduced with group graphs. The activities can be self-paced and arranged in order from the simplest to most complex, but should be used only after students have had many group graphing experiences.

GRAPH-IT BAGS

Graph-it bags are an easily assembled tool that provides independent graphing experiences at the real graph stage. Graph-it bags are filled with objects that students can classify by an attribute such as size, shape, color, or kind.

Place six to ten graph-it bags, each containing different objects, at a center. A student takes a graph-it bag and a reusable grid, determines the common attribute by sorting and classifying (this could be different from your original intent and still be valid), and arranges the objects by attribute, one object to each square. For example, one bag might contain plastic cars and trucks of different colors. Students could choose to sort the toys by type—car or truck—or by color. Students could sort dyed pasta by color or shape.

A series of graph-it bags could range from lower- to higher-level graphing experiences. Ideally, a set would include separate bags containing objects that could be sorted by two, three, four, five, or six attributes. The objects with just two significant attributes, such as large and small, would provide the simplest graphing experiences. As students are ready, they can move to bags with more attributes, such as five different types of pasta. By providing a range, you are assured that all students have the opportunity to be challenged.

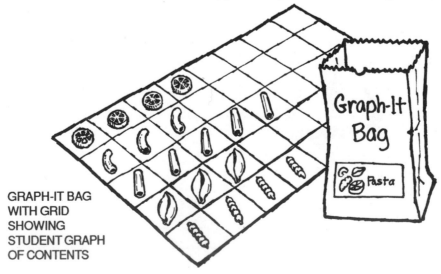

GRAPH-IT BAG
WITH GRID
SHOWING
STUDENT GRAPH
OF CONTENTS

Objects for graph-it bags can be inexpensive materials, readily available, such as the following:

blocks	safety pins	bottle caps
buttons	rocks	postcards
shells	keys	toy animals
beans	seeds and pods	toy vehicles
leaves	crayons	fruit pits
pasta	nuts and bolts	bread tags
fabric scraps	game pieces	marbles

Graph-it bags can be made from a variety of containers, including lunch bags, zip-lock bags, cloth bags with drawstrings, or even margarine tubs or small boxes. Label them on the outside with both the name and a picture of the objects enclosed. The blackline master on page 26 gives you labels for 16 different graph-it bags.

You will need reusable grids for students to use with the graph-it bags. *Graphing Grids* (from Dale Seymour Publications) work well. You could make your own reusable grids by ruling 3-inch squares on a piece of tagboard and covering it with clear contact paper or laminating film for durability. If the objects you want to graph are larger than 3 inches, measure them and rule a larger grid. For graphing objects smaller than one inch, copy the blackline masters on pages 27–30.

After graphing, the student should express—to you, a parent helper, a cross-age tutor, or another student—three statements about what he or she notices about the graph. For example, "There are more macaroni pieces than pasta wheels. There are fewer pasta shells than pasta spirals. There are the same number of elbow macaroni as pasta shells."

After expressing three statements about their graphs, students can use worksheets like the one provided on page 31 to make their own permanent records of the graph-it bags at the pictorial or abstract level. Ask students to write down their conclusions as an extension of this activity. Graph-it bags are excellent for individualizing instruction as they allow students to work at either the concrete, oral, or abstract written level.

LABELS FOR GRAPH-IT BAGS

	Beans		Pattern blocks
	Pasta		Sequins
	Bread tags		Buttons
	Seeds		Paper clips
	Rocks		Game pieces
	Nuts and bolts		Keys
	Unifix cubes		Beads
	Caps		Shells

REUSABLE GRID FOR TWO ATTRIBUTES

REUSABLE GRID FOR THREE ATTRIBUTES

REUSABLE GRID FOR
FOUR ATTRIBUTES

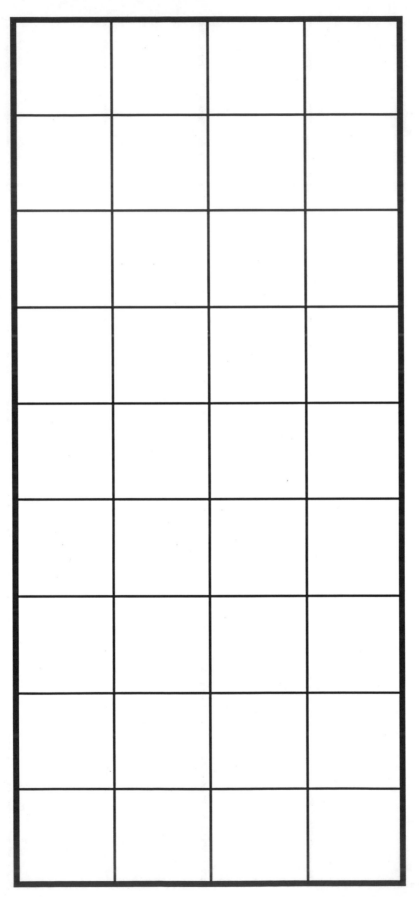

REUSABLE GRID FOR FIVE ATTRIBUTES

REUSABLE GRID FOR SIX ATTRIBUTES

RECORD OF MY GRAPH-IT BAG

Make a permanent record of your graph-it bag.

Which row has the most? _____

Which row has the least? _____

Write a statement about your graph.

GRAPHING FOLDERS

Graphing folders are a good way to provide independent experiences at the picture and abstract stages of graphing. Each graphing folder has a picture on the front and copies of student worksheets inside. The picture contains various numbers of objects for students to count, then graph.

This activity was designed for use at two levels. At Level I, grades K–1, students simply count the objects and record their findings on the grid, then discuss their results with the teacher or an aide. At Level II, grades 1–2, students go on to answer written questions on sheets they find inside the folder.

On pages 34–61 you will find the materials for assembling 14 different graphing folders. Copy them, cut apart the large picture and the grid below it, and mount the picture on a manila file folder. Inside the folder, place copies of the grid along with copies of the question sheets for more advanced students. These folders can be kept at a center for independent use, or you can have all the students in a class working on the same page together.

SAMPLE ILLUSTRATION FOR
GRAPHING FOLDER

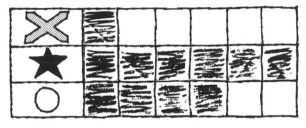

GRID WITH CORRESPONDING
ILLUSTRATIONS

You will need to provide oral directions before launching this activity. For Level I, you need to be sure students understand how to record their findings on the grid. For the sample folder "Flag," your directions might be as follows:

Use the picture to complete the graph.
Count the crosses. Color one square to show one cross.
Count the stars. Color six squares to show six stars.
Count the circles. Color an equal number of squares.

After students have finished a graph, encourage them to discuss and interpret it. Begin by asking for direct information with questions that involve simple counting or comparison: How many? Which has least? or Which has most? Then ask for higher-level information with questions that involve making calculations: How many more? How many less? or What is the total of all the things?

Level II students will do the same type of follow-up in writing, using the question sheet they find inside each folder. Note that although answers are to be written, little or no reading is required. Again, you will need to provide oral directions at first to be sure students know how to respond to the question sheets. In particular, be sure they recognize the symbols that cue the words *more/most* and *fewer/least*. The example that follows, based on the folder "Flag," shows how you might lead students through their first question sheet.

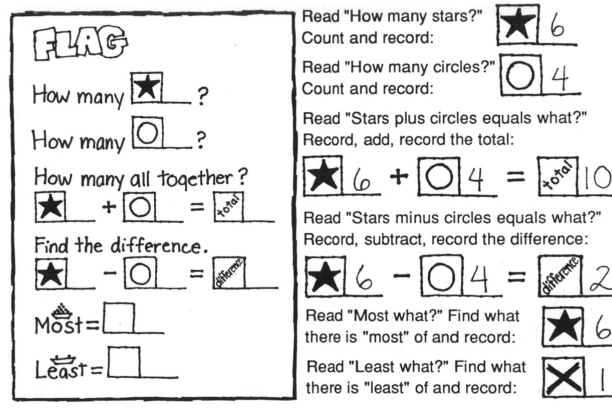

The graphing folder activities in this book are organized according to number of attributes, beginning with two and increasing to six. On pages 62–65 you will find blank forms for creating additional graphing folders, one with space for three attributes, another with space for four. You might have students draw their own pictures, take photographs, or locate appropriate magazine illustrations. Caution students to avoid pictures that show only part of an object, such as half an apple. An open-ended question sheet accompanies each blank form; note that illustrations are to be added, corresponding to the student's own graphing picture (this step is to be completed by the teacher).

FROG POND

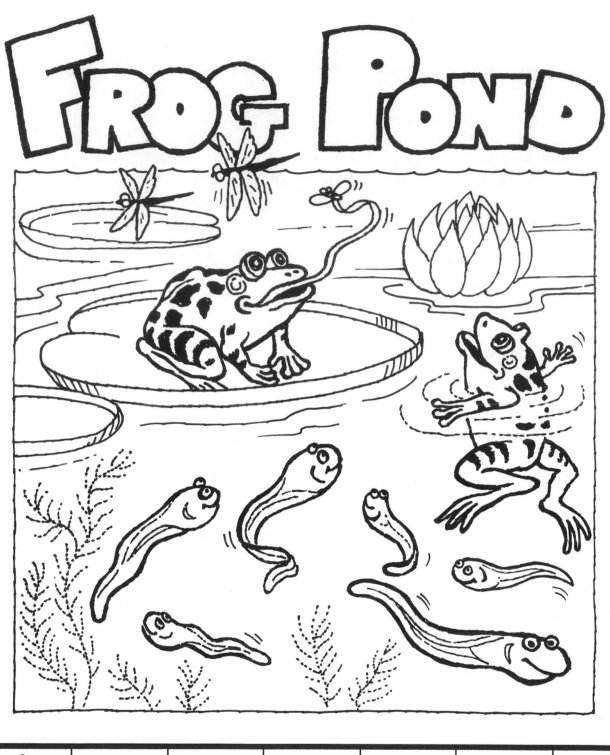

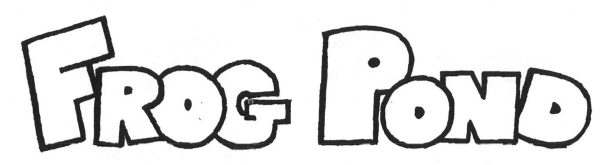

How many  _____ ?

How many  _____ ?

How many all together?

 _____ + 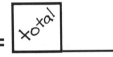 _____ = total _____

There are more ☐ _____ .

There are fewer ☐ _____ .

Clown Capers

Clown Capers

How many _____?

How many _____?

How many all together?

_____ + _____ = _____

There are more ☐_____.

There are fewer ☐_____.

SPIDER PLANT

How many ⬜ _____ ?

How many ⬜ _____ ?

How many all together?

⬜ _____ **+** ⬜ _____ **=** ⬜ _____

There are more ⬜ _____ .

There are fewer ⬜ _____ .

Find the difference.

⬜ _____ **—** ⬜ _____ **=** ⬜ _____

FISH BOWL

How many  _____ ?

How many _____ ?

How many _____ ?

How many all together?

_____ + _____ + _____ = total _____

_____ + _____ = total _____

Most = ☐ _____

Least = ☐ _____

How many ⬛🥾 ____?

How many ⬛🎩 ____?

How many ⬛☂ ____?

How many all together?

⬛🥾 ____ + ⬛🎩 ____ + ⬛☂ ____ = [total] ____

⬛☂ ____ + ⬛🎩 ____ = [total] ____

Most = ▢ ____

Least = ▢ ____

BEE HIVE

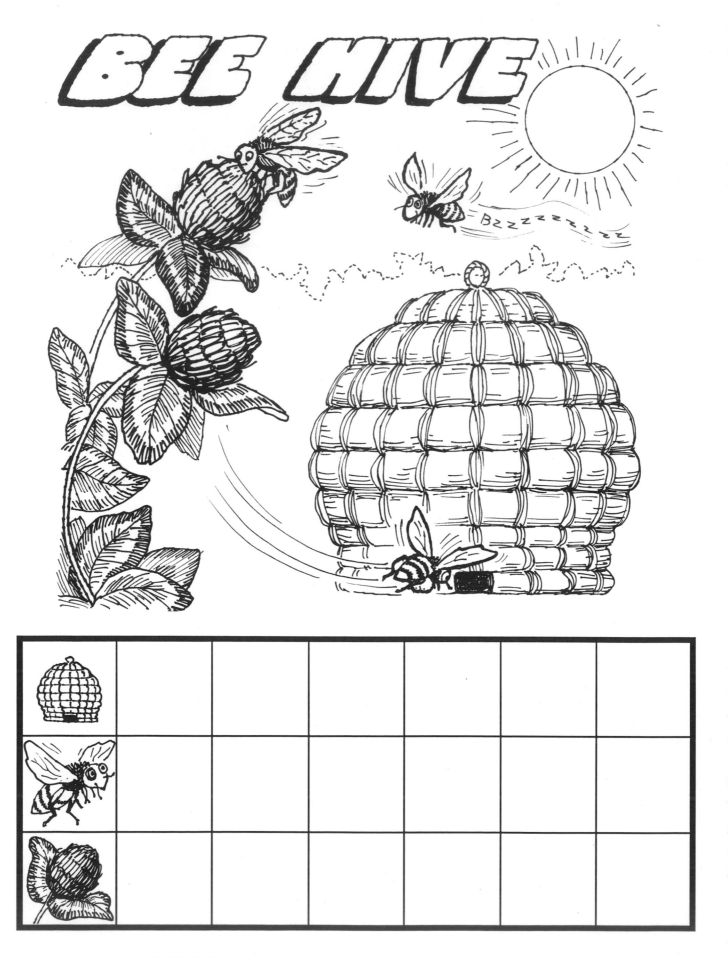

How many 🏠 _____ ?

How many 🐝 _____ ?

How many 🌸 _____ ?

How many all together?

🏠 _____ + 🐝 _____ + 🌸 _____ = ᵗᵒᵗᵃˡ _____

There are more _____ .

There are fewer _____ .

Find the difference.

🐝 _____ – 🌸 _____ = ᵈⁱᶠᶠᵉʳᵉⁿᶜᵉ _____

KITES

How many 🪁 _____?

How many 🎏 _____?

How many 🪁 _____?

How many 🐛 _____?

How many all together?

🐛 _____ + 🪁 _____ + 🪁 _____ + 🎏 _____ = total _____

Flower Pot

Flower Pot

How many 🌼 _____ ?

How many 🍃 _____ ?

How many 🌹 _____ ?

How many 🪴 _____ ?

How many all together?

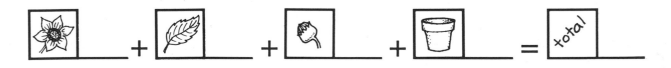

Most = ☐ _____

Least = ☐ _____

Night Sky

✳						
☄						
☆						
☾						

Night Sky

How many ☾ _____ ?

How many ✳ _____ ?

How many ☄ _____ ?

How many ☆ _____ ?

Find the total.

☾ _____ + ☆ _____ + ☄ _____ + ✳ _____ = total _____

Find the difference.

☆ _____ − ☄ _____ = difference _____

acorn						
squirrel						
ladybug						
owl						
tree						

Forest Friends

How many 🌰 _____?

How many 🐿️ _____?

How many 🐞 _____?

How many 🦉 _____?

How many 🌳 _____?

Add. 🌰 _____ + 🐿️ _____ = _____

Add. 🐞 _____ + 🦉 _____ = _____

The Castle

How many ⌒ _____ ?

How many △ _____ ?

How many ▭ _____ ?

How many ◻ _____ ?

How many ▱ _____ ?

Add. ▱ _____ + △ _____ = _____

Subtract. ▭ _____ − △ _____ = _____

Add. ▱ _____ + ⌒ _____ = _____

Subtract. ▱ _____ − ⌒ _____ = _____

Pattern Blocks

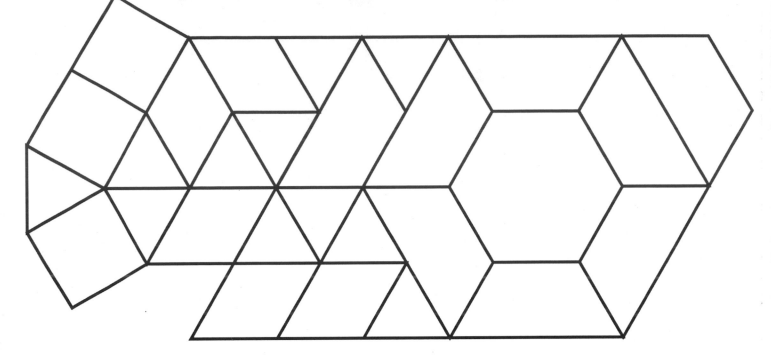

Pattern Blocks

How many ⬓ _____?

How many ◇ _____?

How many ⬡ _____?

How many □ _____?

How many △ _____?

Find the total.

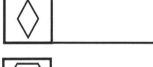

□ _____

+ △ _____

_____

_____

Find the difference.

⬓ _____

− ⬡ _____

_____

_____

Dinnertime

How many [🌽] _____ ?

How many [○] _____ ?

How many [🥔] _____ ?

How many [🍖] _____ ?

How many [🫛] _____ ?

How many [🥛] _____ ?

[🫛] _____ + [🥔] _____ = _____

[○] _____ − [🥛] _____ = _____

[🍖] _____ + [🫛] _____ + [○] _____ = _____

Gingerbread House

How many [🍪] _____?

How many [🌽] _____?

How many [♡] _____?

How many [🍊] _____?

How many [🍬] _____?

How many [🍭] _____?

[🍪] _____ + [🍬] _____ = _____

[🍊] _____ + [♡] _____ = _____

How many more [🌽] than [♡] ?

[🌽] _____ − [♡] _____ = _____

My Graphing Picture

Draw a graphing picture.

Graph your information.

Tell the student to draw a picture with varying numbers of three different objects.
Then have the students graph the information.

My Graphing Picture

How many ☐ ☐_____?

How many ☐ ☐_____?

How many ☐ ☐_____?

How many all together?

☐ ☐_____ + ☐ ☐____ + ☐ ☐____ = ☐total ☐_____

☐ ☐_____ + ☐ ☐____ = ☐ ☐____

☐ ☐_____ − ☐ ☐____ = ☐ ☐____

Most = ☐ ☐_____

Least = ☐ ☐_____

Add small versions of the students' own pictures in the boxes to make questions for their graphing pictures.

My Graphing Picture

Draw a graphing picture.

Label the objects here.

Give this to a friend to complete.

Tell the student to draw a picture with varying numbers of five different objects. Have the student label the objects and then ask a friend to graph the information.

My Graphing Picture

How many ☐____?

How many ☐____?

How many ☐____?

How many ☐____?

Find the total.

☐____ + ☐____ + ☐____ = [total]____

🏆 Most = _____

👑 Least = _____

Add small versions of the students' own pictures in the boxes to make questions for their graphing pictures.

INDIVIDUAL POLLING GRAPHS

Individual polling graphs are another good way to provide independent experiences at the picture and abstract stages of graphing. Students can use these graphs to discover information about each other, such as the most common month for birthdays, favorite pets, the most common first letter of their names, favorite colors, and the average number of people in their families. On pages 67–76 you will find blackline masters for both specific and open-ended polling graphs. Copy them and ask students to complete these graphs either at school or at home. At school, students can poll their peers in the classroom or poll students from other classes on the playground. At home, students can poll their family, friends, or neighbors. As each person polled answers the question, students mark the responses on individual grids. They answer questions about the graph when it is complete.

The best results occur when students develop their own questions to investigate their personal interests or to discover information that can help solve real problems. For instance, if students were concerned that the food served in the cafeteria was not what they wanted to eat, they could speak to cafeteria personnel and discover if other choices were available. They could poll the school population about favorite and least favorite food and share the results with cafeteria personnel. If students find that tacos are more popular than hamburgers, perhaps tacos could be served each Wednesday. If students prefer green salad to green beans, perhaps salad could be on the menu every day.

If the number of squares on a grid is not enough to accommodate the responses, encourage students to figure out different ways of solving the problem and discuss these solutions in groups.

COMPLETED
POLLING GRAPH

When is your birthday?

January								
February								
March								
April								
May								
June								
July								
August								
September								
October								
November								
December								

Do you prefer oranges or apples?

orange												
apple												

How many   🍊 ?   There are more   ☐ .

How many   🍎 ?   There are fewer   ☐ .

How did you come to school?

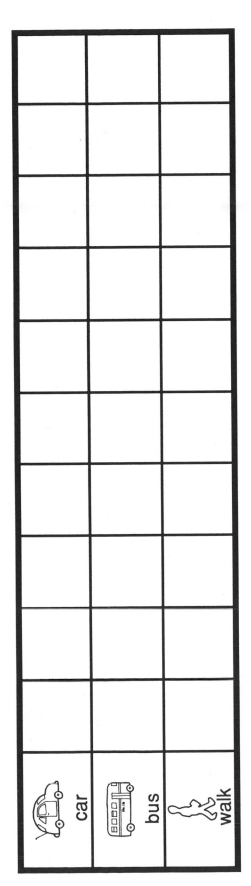 car											
bus											
walk											

Find the total.

How many 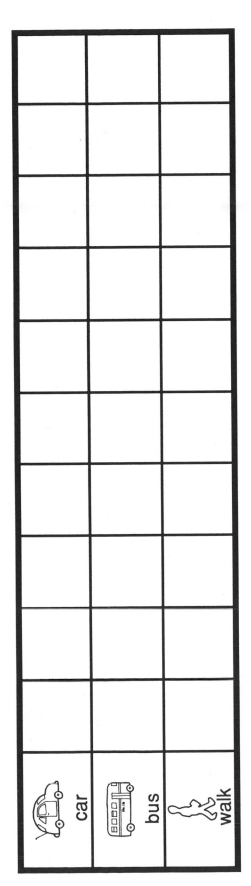 ? _____

How many ? _____

How many ? _____

 +  + 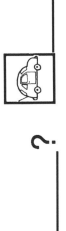 = total

Most = ☐

Least = ☐

What color is your hair?

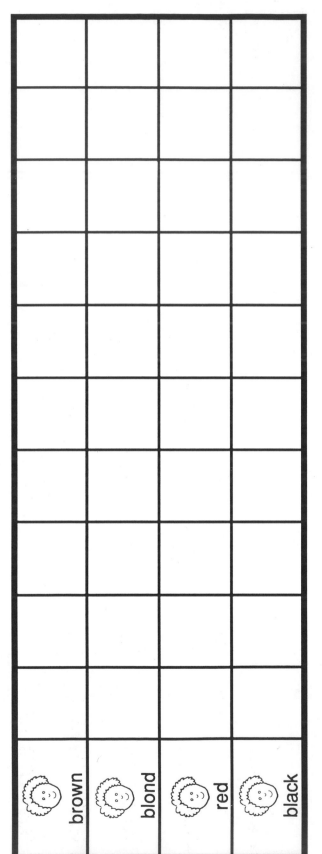

	brown
	blond
	red
	black

How many **?** ▲ Most = ☐

How many **?** ☺ Least = ☐

How many **?**

How many **?**

Find the total.

☐ brown + ___ ☐ blond + ___ ☐ red + ___ ☐ black = ___ total

What is your favorite pet?

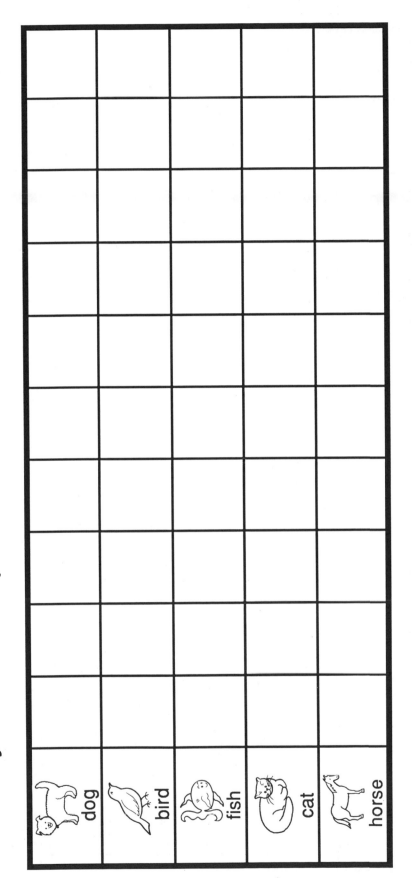

🐕 dog										
🐦 bird										
🐠 fish										
🐈 cat										
🐴 horse										

How many 🐕 ? _____

How many 🐦 ? _____

How many 🐠 ? _____

How many 🐈 ? _____

How many 🐴 ? _____

There are more _____ .

There are fewer _____ .

_____ + _____ = [total]

What color are your eyes?

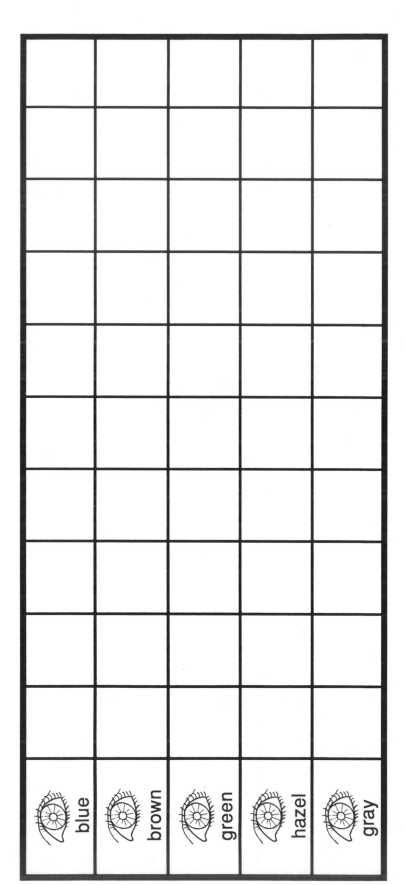

👁 blue										
👁 brown										
👁 green										
👁 hazel										
👁 gray										

How many [blue] ? _____

How many [brown] ? _____

How many [green] ? _____

How many [hazel] ? _____

How many [gray] ? _____

There are more _____ .

There are fewer _____ .

[blue] + [hazel] + [green] = [Total]

Polling Question _____

Find the total.

How many □ ?

How many □ ?

There are more ▲ □ .

There are fewer ▽ □ .

□
+
□
_____
[total]

Polling Question _____

How many _____?

How many _____?

How many _____?

🔺 Most = ☐

😊 Least = ☐

How many all together?

☐ + ☐ + ☐ = ☐ total

Polling Question _____

[A 4-column by 11-row grid/graph]

How many [ ] ?
How many [ ] ?
How many [ ] ?
How many [ ] ?

[ ] + [ ] = [total]
[ ] + [ ] = [total]

How many all together?

[ ] + [ ] + [ ] + [ ] = [total]

Polling Question _____

There are more ◀ _____ .

There are fewer ⬆ _____ .

How many ? [ ] [ ]

How many ? [ ] [ ]

How many ? [ ] [ ]

How many ? [ ] + [ ] + [ ] = [total]

How many ? [ ]

GATHERING DATA FROM HOME

Students can help gather data to use for classroom graphing experiences by bringing in counts of various items in their homes. This helps ensure that the information being graphed is relevant to their lives. Use this approach after you have introduced graphing in the classroom so that home participation is a follow-up of school activities. On the following pages you will find counting slips, tally sheets, counting booklets, and individual inventories for gathering data from home.

Counting slips (pages 78–80). Make enough copies of each blackline master for each class member. Cut each sheet into eight slips and distribute one slip at a time so that each student counts the same object on any given day. For example, give all the students the slip with the hand on it and ask them to take it home and count all the fingers in their homes. Students count the specified objects, in this case fingers, and either write the number they find or make tally marks. When they bring their recording sheets back the next day, have them graph and discuss the results. Used in this way, each blackline master provides eight homework sessions.

Tally sheets (pages 81–84). Give each student one copy of a tally sheet to take home. When counting objects, such as windows in the house, the student will not be able to see all of them at once and can make tally marks to keep count. The students bring their recording sheets back the next day and graph and discuss the results.

Counting booklets (pages 85–87). Cut these sheets into four sections and assemble them into booklets. Use the booklet as long-term homework. For example, students might take a week to finish all the counting activities in the booklet. Each student brings his or her completed booklet back to school. Use the counting results as data for a group graph.

Individual inventories (pages 88–91). On these sheets, the students not only count objects found at home (for example, in the refrigerator), they also record these findings on grids and answer questions comparing the reults. Use the results from the students' graphs as data for a group graph.

A COMPLETED COUNTING SLIP

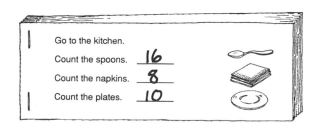

A COMPLETED COUNTING BOOKLET

Counting slip

Name _____

Count the telephones.

Counting slip

Name _____

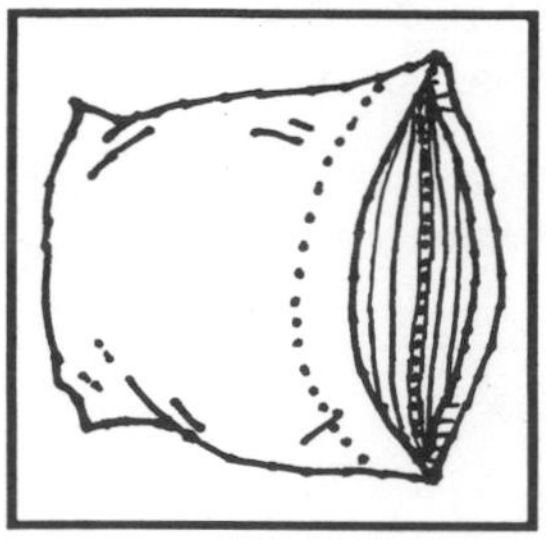

Count the pillows.

Counting slip

Name _____

Count the purses.

Counting slip

Name _____

Count the tables.

Counting slip

Name _____

Count the ties.

Counting slip

Name _____

Count the chairs.

Counting slip

Name _____

Count the pens.

Counting slip

Name _____

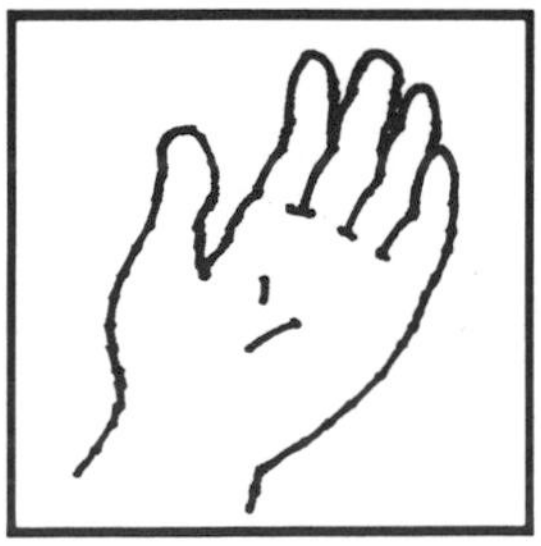

Count the fingers.

Counting slip

Name _____

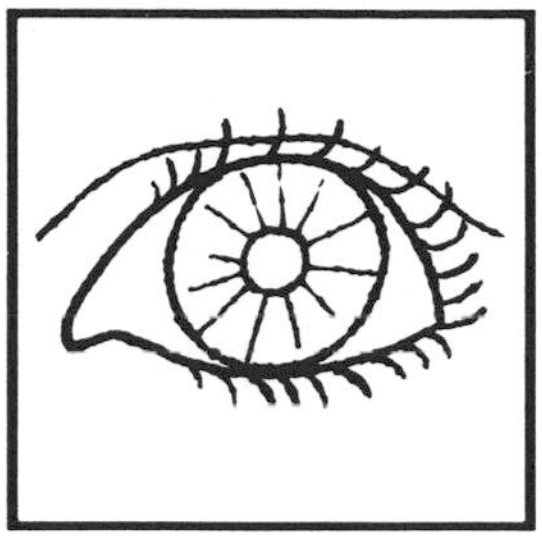

Count the eyes.

Counting slip

Name _____

Count the pets.

Counting slip

Name _____

Count the stuffed animals.

Counting slip

Name _____

Count the clocks.

Counting slip

Name _____

Count the beds.

Counting slip

Name _____

Count the faucets.

Counting slip

Name _____

Count the radios.

Counting slip

Name _____

Count the mirrors.

Counting slip

Name _____

Count the watches.

Counting slip

Name _____

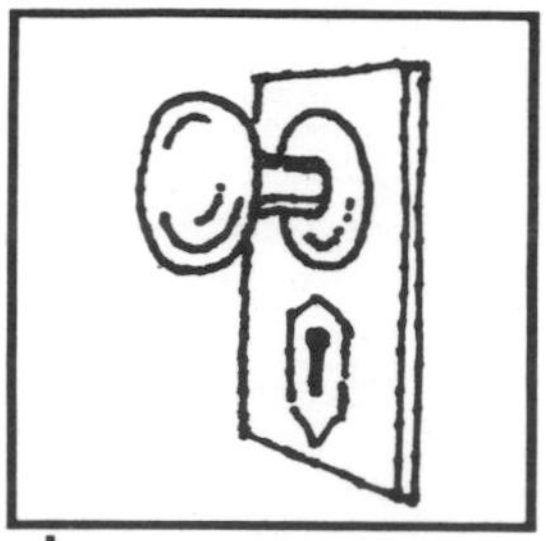

Count the door knobs.

Counting slip

Name _____

Count the televisions.

Counting slip

Name _____

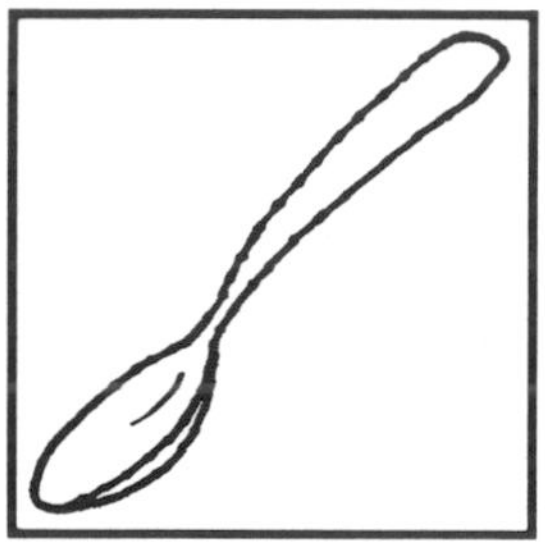

Count the spoons.

Counting slip

Name _____

Count the sweaters.

Counting slip

Name _____

Count the house plants.

Counting slip

Name _____

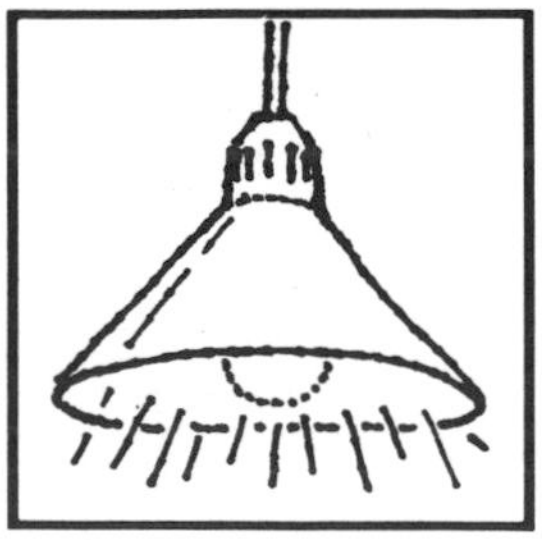

Count the lights.

Counting slip

Name _____

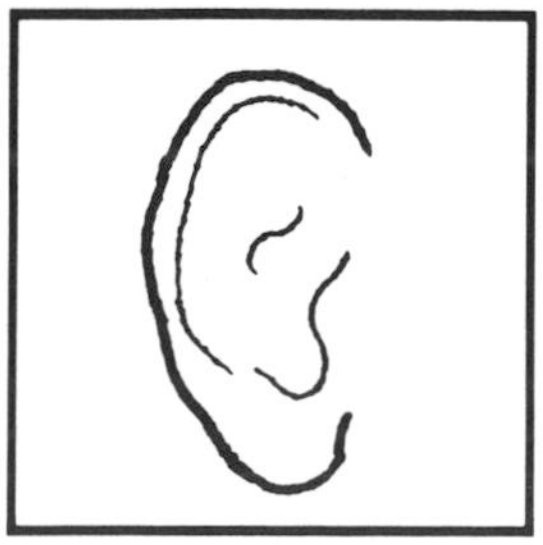

Count the ears.

Name _____

Count the windows in your house.

Make your tally marks.

Circle each group of 10 tally marks.

How many groups of 10?

How many ones?

How many windows?

Name _____

Count the shoes in your house.

Make your tally marks.

Circle each group of 10 tally marks.

How many groups of 10?

How many ones?

How many shoes?

Name _____

Count the light switches in your house.

Make your tally marks.

Circle each group of 10 tally marks.

How many groups of 10?

How many ones?

How many light switches?

Name _____

Count the books in your house.

Make your tally marks.

Circle each group of 10 tally marks.

How many groups of 10?

How many ones?

How many books?

Counting booklet

Go to the kitchen.

Count the forks. ________

Count the cups. ________

Count the pans. ________

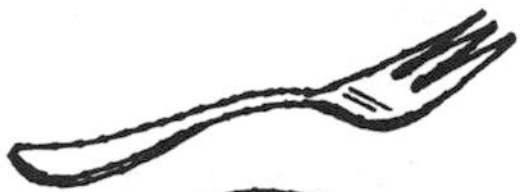
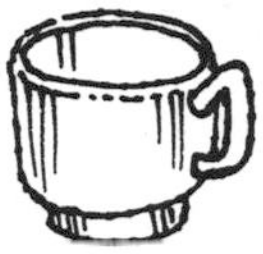
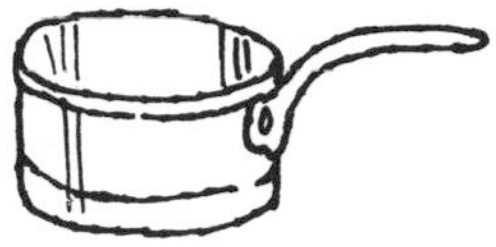

Go to the bathroom.

Count the faucets. ________

Count the towels. ________

Count the soap. ________

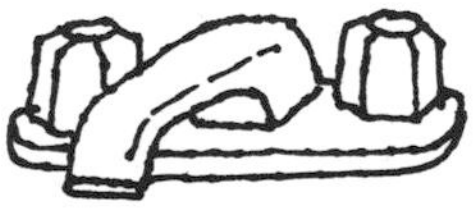
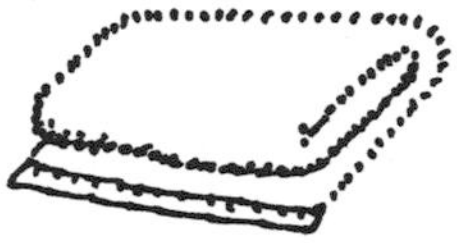
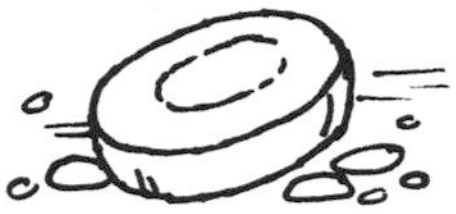

Go to the bedroom.

Count the pillows. ________

Count the drawers. ________

Count the books. ________

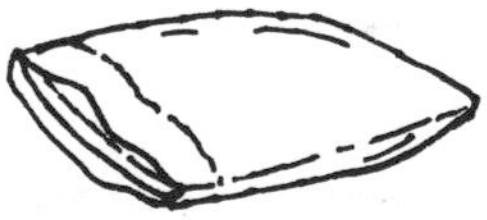
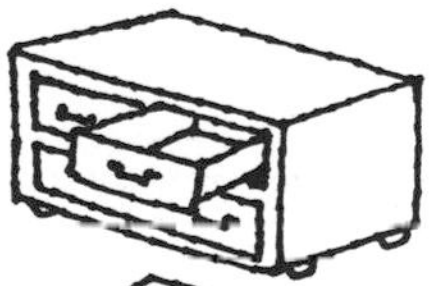
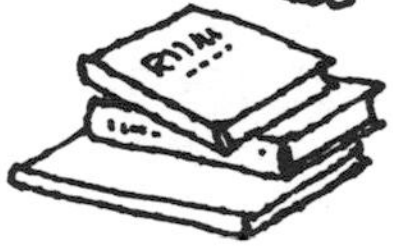

Go to the living room.

How many clocks? ________

How many house plants? ________

How many mirrors? ________

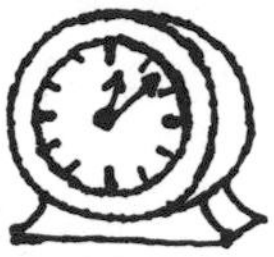
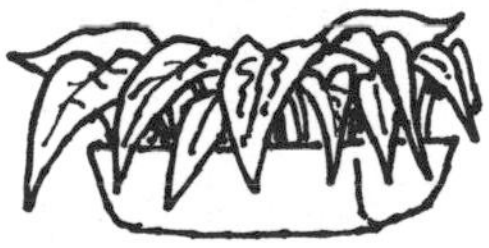
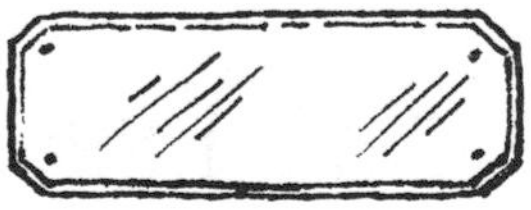

Go to the bathroom.

Count the Band-Aids. _____

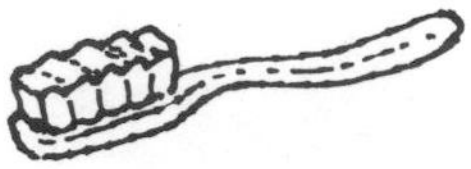

Count the toothbrushes. _____

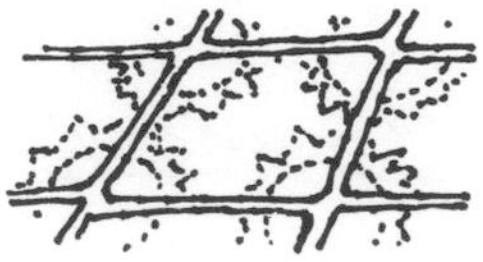

Count the tiles. _____

Go to the kitchen.

Count the spoons. _____

Count the napkins. _____

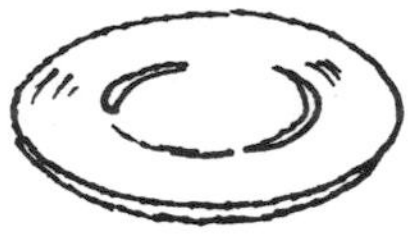

Count the plates. _____

Go to the living room.

Count the televisions. _____

Count the chairs. _____

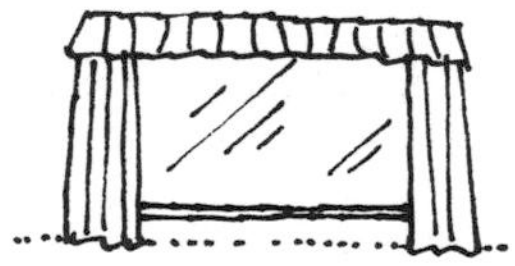

Count the windows. _____

Look at your toys.

Count the games. _____

Count the balls. _____

Count the vehicles. _____

Go to the kitchen cupboard.

Count the bottles. _____

Count the boxes. _____

Count the cans. _____

Go to the living room.

Count the drapes. _____

Count the tables. _____

Count the lamps. _____

Go to the closet.

Count the coats. _____

Count the shoes. _____

Count the belts. _____

Go to the bedroom.

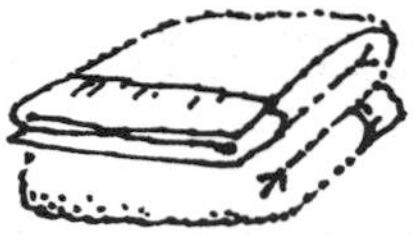

Count the blankets. _____

Count the rugs. _____

Count the pictures. _____

Graph your cereal boxes.

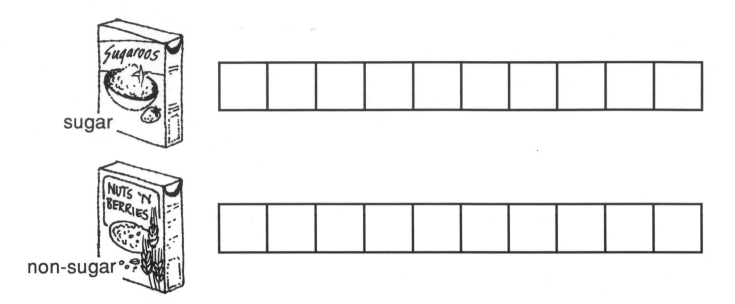

sugar

non-sugar

How many ⬛ sugar ____ ?

How many ⬛ non-sugar ____ ?

More ▢ ____

Less ▢ ____

What is the total?

sugar ____ + non-sugar ____ = total ____

Graph the pictures on your walls.

photographs

paintings

other

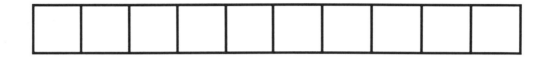

How many _____?

How many _____?

How many _____?

 Most = ☐ _____

 Least = ☐ _____

Write a number sentence.

GRAPHING PRIMER   **89**

Look in the refrigerator. Graph what you see.

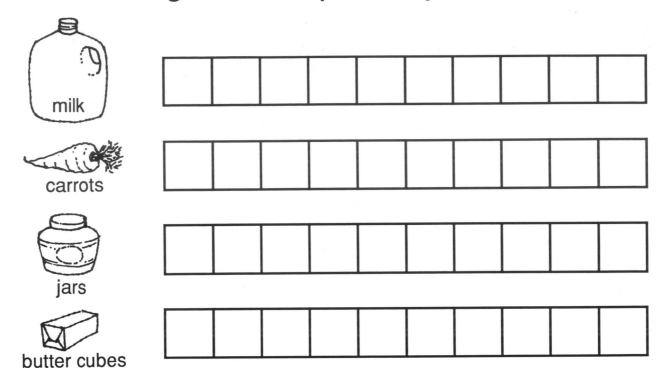

milk

carrots

jars

butter cubes

How many  ?    Most = ⬜ _____

How many  ?   Least = ⬜ _____

How many  ?

How many ▱ ?

Write a number sentence.

Graph your toys.

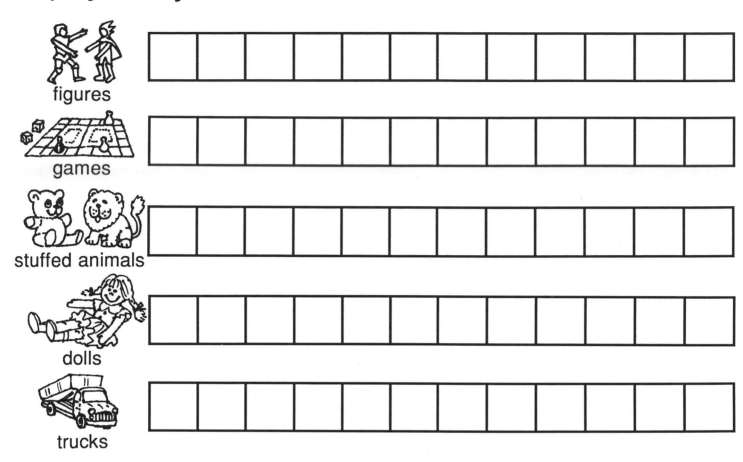

figures

games

stuffed animals

dolls

trucks

How many _____? Most = [ ]_____

How many _____? Least = [ ]_____

How many _____?

How many _____?

How many _____?

Write a number sentence.

Graphing Expert

_____

has demonstrated the ability to
make and interpret simple graphs.

_____   _____

Teacher   Date

Graphing Expert

_____

has demonstrated the ability to
make and interpret simple graphs.

_____   _____

Teacher   Date